100

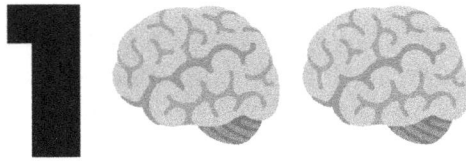

MIND-BLOWING
HUMAN BODY
FACTS

100 Astonishing Wonders of the Body You Never Knew

FELIX GRAYSON

MINDSPARK
PUBLISHING

CONTENTS

BEFORE WE DIVE IN...

Did you know that this is just **one** of many **mind-blowing** books waiting to be discovered?

What if I told you there's a **world of jaw-dropping, unbelievable, and downright bizarre facts** across **sports, science, history, mysteries, and more**—each one packed with stories that will **challenge what you thought you knew?**

EVER WONDERED WHAT IT'S LIKE TO...

- Witness **record-breaking Olympic moments** that defy human limits?

- Explore **real-life conspiracy theories** that sound too wild to be true?

- Discover **unsolved mysteries** that still leave experts baffled?

- Learn about **billionaires, stock market**

crashes, and money secrets?

- Find out how **robots, AI, and space travel are shaping the future?**

- Experience the **most extreme sports, legendary battles, and shocking events?**

This is just the beginning. The **100 Mind-Blowing series** covers it **all**.

WANT TO SEE WHAT'S NEXT?

Go to **FelixGrayson.com** and explore the **growing collection** of books and audiobooks that will **entertain, amaze, and keep you coming back for more.**

FelixGrayson.com

Curiosity doesn't stop here—this is just the beginning. What will blow your mind next?

INTRODUCTION

Welcome to *100 Mind-Blowing Human Body Facts*, a collection designed to make you say, "Wait, seriously? That happens inside me?" From jaw-dropping biological quirks to hidden superpowers you didn't know you had, this book is packed with stories that will make you look at yourself in a whole new way.

Have you ever wondered why your fingers wrinkle in water? Or how your body glows—even though you can't see it? What about the fact that you can survive without a pulse, or that your bones are stronger than steel? These are just a few of the unbelievable truths waiting for you inside. Each fact has been carefully chosen to surprise, entertain, and maybe even make you pause in awe at what your body is doing right now.

Whether you're here for a dose of weird science, a fun party fact, or just want to get to know your own body a little better, this book has you covered. Read it cover to cover, or flip to a random page and let curiosity be your guide. There's no right or wrong way to explore the strange, surprising, and absolutely fascinating world you live in—your own body.

So find a cozy spot, settle in, and get ready to be amazed by the incredible machine that is *you*. Who knows? By the end, you might have a whole new appreciation for what's going on beneath your skin. Let's dive in!

Mind-Blowing Human Body Fact #1

YOUR STOMACH HAS A SECRET SKIN

Every few days, your stomach regenerates its entire lining—because if it didn't, it would digest itself!

That's right: the acid in your stomach is strong enough to dissolve metal. It's called hydrochloric acid, and it's so corrosive that if it came into contact with other parts of your body, it would cause serious damage. To prevent this acid from destroying your insides, your stomach has evolved to constantly rebuild its protective lining every 3 to 4 days. It's like having a built-in renovation crew that never takes a break!

Mind-Blowing Human Body Fact #2

YOU GLOW IN THE DARK

Humans actually emit a faint glow—but it's 1,000 times too dim for our eyes to see.

This eerie phenomenon is called **bioluminescence**, and it's caused by chemical reactions in our cells as they metabolize. Scientists in Japan captured this invisible glow using ultra-sensitive cameras and discovered that our bodies shine brightest in the late afternoon, especially around the cheeks, forehead, and neck. So yes—technically speaking, you're a glowing creature of light… just not bright enough to ditch your flashlight.

Mind-Blowing Human Body Fact #3

BONES ARE STRONGER THAN STEEL

Ounce for ounce, human bone is stronger than concrete—and even some types of steel!

The femur, your thigh bone, can withstand about **30 times** your body weight in force. In terms of strength-to-weight ratio, bone outperforms construction-grade steel. What makes this even more mind-blowing is how lightweight and flexible bone is—allowing it to absorb impact without snapping. So the next time someone calls you "fragile," remind them you're basically made of high-tech scaffolding.

Mind-Blowing Human Body Fact #4

YOU'RE TALLER IN THE MORNING

When you wake up, you're actually about half an inch taller than when you go to bed.

Throughout the day, gravity compresses the soft discs between your spine's vertebrae as you stand, walk, and sit. This subtle squishing causes your height to shrink ever so slightly by nighttime. But while you sleep, your spine stretches back out like a relaxed spring, restoring your full height by morning. So technically, you're a little taller every time you hit snooze!

Mind-Blowing Human Body Fact #5

YOUR BRAIN RUNS ON ELECTRICITY

Your brain generates enough electricity to power a light bulb.

At any given moment, there are around **86 billion neurons** in your brain firing electrical signals to keep your thoughts, memories, movements, and emotions going. These signals generate about **20 watts** of electrical power—roughly the same as a dim household bulb. So if you've ever said you were "burnt out," you weren't entirely speaking metaphorically. You're literally a walking, thinking power plant!

Mind-Blowing Human Body Fact #6

YOU SHED A LOT OF SKIN

Every minute, your body sheds about **30,000 to 40,000** skin cells.

That adds up to roughly **9 pounds** of dead skin every year! Your outermost layer of skin, the epidermis, is constantly regenerating to protect you from the outside world. In fact, every **month or so**, you get a brand-new epidermis. Most of that old skin becomes household dust—so yes, part of what you're cleaning off your shelves is... you.

Mind-Blowing Human Body Fact #7

YOUR HEART CAN BEAT OUTSIDE YOUR BODY

Your heart has its own electrical system—so it can beat on its own.

As long as it has oxygen, the heart doesn't need signals from the brain to keep pumping. In fact, doctors have observed hearts continuing to beat **even after being removed from the body** during transplant procedures. That's because it generates its own rhythm through a group of cells called the **sinoatrial node**, also known as the heart's natural pacemaker. It's like a drummer that never misses a beat—even when playing solo.

Mind-Blowing Human Body Fact #8

YOUR TONGUE HAS A UNIQUE PRINT

Just like your fingerprints, your tongue has a one-of-a-kind pattern.

No two tongues are exactly alike—each has its own unique shape, texture, and distribution of taste buds and grooves. Scientists have even explored the idea of using **tongue prints** for biometric identification, since they're so distinct and hard to replicate. So while it might not replace your fingerprint anytime soon, your tongue is quietly carrying its own secret signature.

Mind-Blowing Human Body Fact #9

YOU HAVE GOLD IN YOUR BODY

The human body contains trace amounts of real gold.

It's true—your blood and tissues naturally contain about **0.2 milligrams** of gold, mostly playing a role in joint health and enzyme function. That's not nearly enough to retire on (sorry!), but it's a fascinating reminder that your body is built from elements scattered across the universe—including precious metals. You're literally worth your weight in stardust... with just a sprinkle of bling.

Mind-Blowing Human Body Fact #10

YOUR LIVER CAN REGROW ITSELF

The human liver is the only organ that can fully regenerate—even if up to **75%** of it is removed.

This amazing ability makes liver transplants and donations possible, where a portion of a healthy liver can be given to someone else and both livers will regrow to near-full size. Ancient myths spoke of regenerating livers (like Prometheus in Greek mythology), and it turns out science wasn't too far behind. Your liver isn't just hardworking—it's a real-life shape-shifter.

Mind-Blowing Human Body Fact #11

YOU CAN LIVE WITHOUT A STOMACH

Believe it or not, you can survive just fine without a stomach.

In extreme medical cases—like cancer or severe ulcers—doctors can perform a **total gastrectomy**, removing the stomach entirely. The esophagus is then connected directly to the small intestine, and with some dietary adjustments, people can continue to eat, digest, and absorb nutrients. It's not ideal, but it's proof of how incredibly adaptable the human body can be. Even without its central digestive chamber, it finds a way to keep going.

Mind-Blowing Human Body Fact #12

YOUR NOSE REMEMBERS 50,000 SCENTS

Your sense of smell is far more powerful than you think—it can identify **around 50,000 different odors**.

The olfactory system is directly connected to the brain's limbic system, which handles memory and emotion. That's why a random whiff of a perfume or food can instantly transport you back to a specific moment in time. While dogs get all the credit for sniffing skills, your nose is quietly cataloging a complex scent library of its own—no training required.

Mind-Blowing Human Body Fact #13

YOUR BLOOD TRAVELS THOUSANDS OF MILES

Every single day, your blood travels about **12,000 miles** through your body—that's roughly halfway around the Earth!

With each beat of your heart, blood is pumped through a vast network of arteries, veins, and capillaries. If you laid out all your blood vessels end to end, they'd stretch over **60,000 miles** in an adult. That's enough to circle the globe **twice and then some**—all inside your body. Your circulatory system is basically the ultimate cross-country traveler.

Mind-Blowing Human Body Fact #14

YOU BLINK MORE THAN YOU BREATHE

On average, you blink about **15–20 times per minute**—which adds up to around **20,000 blinks a day**.

That's more than the number of breaths you take daily! Blinking keeps your eyes moist, clears away dust, and even gives your brain tiny micro-breaks. In fact, you spend about **10% of your waking hours** with your eyes closed just from blinking. It's an invisible rhythm your body keeps—like a built-in windshield wiper for your vision.

Mind-Blowing Human Body Fact #15

YOUR MUSCLES OUTNUMBER YOUR BONES

The human body has **more than 600 muscles**—nearly **three times** the number of bones.

While you have 206 bones in your adult skeleton, your muscular system is far more complex. Muscles control everything from massive movements like running to tiny ones like blinking or moving your eyes. Even your facial expressions rely on over **40 muscles** working in perfect coordination. So next time someone calls you a "bag of bones," remind them you're mostly muscle.

Mind-Blowing Human Body Fact #16

YOUR EARS NEVER STOP HEARING

Even when you're asleep, your ears stay switched on.

Your brain may tune out the noise, but your ears are still actively receiving sound. That's why loud noises—like alarms, thunder, or someone calling your name—can jolt you awake. Unlike your eyes, which get to shut down completely, your ears are always listening, 24/7. They're the night shift workers of your sensory team, standing guard while you dream.

Mind-Blowing Human Body Fact #17

YOU'RE WIRED WITH 37 TRILLION CELLS

Your body is made up of an estimated **37 trillion** cells—all working together in perfect harmony.

These microscopic units form your skin, bones, blood, brain, and everything in between. What's even more mind-blowing? Each of those cells contains **around 6 feet of DNA**, tightly coiled into a microscopic nucleus. If you unraveled all the DNA in your body, it could stretch to **Pluto and back**—multiple times. You're not just a person—you're a walking universe of living machinery.

Mind-Blowing Human Body Fact #18

GOOSEBUMPS ARE A SURVIVAL REFLEX

Those tiny bumps on your skin? They're an ancient defense mechanism.

Goosebumps happen when tiny muscles at the base of hair follicles contract—usually in response to cold or strong emotions. But originally, they served a more primal purpose: **to puff up your body hair and make you look bigger to predators** (or to trap heat). While modern humans don't have enough body hair for this to work, the reflex stuck around as a quirky biological souvenir.

Mind-Blowing Human Body Fact #19

YOUR BODY HEATS LIKE A POWER PLANT

In just 30 minutes, your body gives off enough heat to boil half a gallon of water.

Thanks to constant chemical reactions—especially in your muscles and organs—your body is a surprisingly powerful heat engine. On average, you radiate around **100 watts** of heat at rest, and much more when you're active. That's why crowded rooms feel warm fast—it's basically a low-key human furnace party.

Mind-Blowing Human Body Fact #20

YOU'RE BORN WITH MORE BONES

Babies are born with about **300 bones**—but adults only have 206.

That's because as you grow, many of those tiny bones gradually fuse together. A perfect example? Your skull. It starts off in multiple flexible pieces to make childbirth easier and protect a growing brain, then hardens into a solid structure over time. So technically, you had more bones as a baby than you do now... which is kind of the opposite of what you'd expect!

Mind-Blowing Human Body Fact #21

YOUR SMALLEST BONE IS IN YOUR EAR

The tiniest bone in your body is called the **stapes**—and it's smaller than a grain of rice.

Located in your middle ear, the stapes (or stirrup bone) plays a huge role in hearing. It vibrates in response to sound waves and passes those signals along to the inner ear. Despite being just **about 0.1 inches long**, it's essential for converting sound into something your brain can understand. Big things really do come in small packages.

Mind-Blowing Human Body Fact #22

YOUR GUT HAS A SECOND BRAIN

Deep in your digestive system is something called the **enteric nervous system**—a complex network of over **100 million neurons** embedded in the walls of your gut.

Scientists often refer to it as your "second brain" because it can operate independently of the brain and spinal cord, managing digestion, signaling discomfort, and even producing neurotransmitters like **serotonin**. That "gut feeling" you get? It's not just a figure of speech—it's your second brain chiming in.

Mind-Blowing Human Body Fact #23

YOUR EYES SEE THE PAST

When you look at anything, you're technically seeing it as it was **in the past**—even if just by a fraction of a second.

Light takes time to travel, even at **186,000 miles per second**. So when it bounces off an object and reaches your eyes, you're seeing a delayed version of reality. For nearby objects, the delay is tiny—mere nanoseconds—but for distant stars, it could be **millions of years**. In other words, your eyes are constantly time-traveling.

Mind-Blowing Human Body Fact #24

YOUR SALIVA CAN FILL A BATHTUB

Over the course of a year, your mouth produces enough saliva to fill a standard bathtub—about **25,000 to 30,000 ounces**!

That's nearly **2 pints a day**, all created to help you digest food, fight germs, and keep your mouth moist. Saliva even contains enzymes that start breaking down food before it reaches your stomach. So while it may not be glamorous, spit is one of your body's most underrated MVPs.

Mind-Blowing Human Body Fact #25

YOUR HAIR IS NEARLY INDESTRUCTIBLE

Human hair is incredibly strong and resistant—it can **withstand more tension than copper wire** of the same diameter.

A single strand can support about **3.5 ounces**, and collectively, your scalp hair could hold the weight of **two elephants**—theoretically, anyway. Even more impressive? Hair doesn't decompose easily. It can survive centuries in archaeological digs, acting as a time capsule of your body's biology long after everything else is gone.

Mind-Blowing Human Body Fact #26

YOUR BODY MAKES ITS OWN DRUGS

Your brain produces natural versions of painkillers, antidepressants, and even a chemical similar to cannabis.

Endorphins act like morphine, blocking pain and creating feelings of pleasure. Dopamine boosts motivation and reward. And **anandamide**—nicknamed the "bliss molecule"—is chemically similar to THC, the active ingredient in marijuana. Your brain is basically a built-in pharmacy, mixing up custom cocktails tailored to your every mood and moment.

Mind-Blowing Human Body Fact #27

YOU CAN TASTE WITHOUT A TONGUE

While the tongue does most of the work, it's not the only part of your body that can detect taste.

Taste receptors are also found on the **roof of your mouth**, the **back of your throat**, and even **inside your nose**. These extra sensors help fine-tune flavor by working together with your sense of smell and texture. So even if you lost your tongue (don't worry—you won't), your body could still recognize the difference between sweet, sour, salty, bitter, and umami.

Mind-Blowing Human Body Fact #28

YOUR BLOOD IS A LIQUID ORGAN

It may look like just fluid, but blood is actually considered a **connective tissue**—making it a full-fledged organ.

Composed of cells suspended in plasma, blood performs dozens of essential functions: carrying oxygen, transporting nutrients, regulating temperature, and fighting off invaders. It even has its own immune system and repair crew. So the next time you see a drop of blood, remember—you're looking at a flowing, living organ in action.

Mind-Blowing Human Body Fact #29

YOU MAKE NEW BLOOD EVERY SECOND

Your body produces about **2 million red blood cells** *every second*.

These cells are created in your bone marrow and live for around **120 days**, constantly cycling through your bloodstream to deliver oxygen and remove carbon dioxide. Over the course of a day, that adds up to a staggering **170 billion** new cells. You're basically a nonstop blood factory—quietly manufacturing life with every heartbeat.

Mind-Blowing Human Body Fact #30

YOUR SKIN CAN SENSE A SINGLE HAIR

Your sense of touch is so precise, you can feel the movement of just **one hair**.

That's because your skin is loaded with specialized receptors that detect pressure, vibration, temperature, and even the slightest shifts on the surface. Ever felt a tiny bug crawl on your arm before seeing it? That's your nervous system responding to incredibly subtle stimuli—faster than conscious thought. You're basically equipped with a full-body motion detector.

Mind-Blowing Human Body Fact #31

YOUR BONES ARE FULL OF HOLES

Despite their toughness, bones aren't solid — they're full of **tiny holes and tunnels**.

Inside, bones are more like a honeycomb than concrete. This spongy structure, called **trabecular bone**, makes them both strong and lightweight. It also houses bone marrow, where new blood cells are made. So while bones might seem like solid sticks, they're actually marvels of biological engineering—hard on the outside, hollow and alive on the inside.

Mind-Blowing Human Body Fact #32

YOU CAN HEAR YOUR EYES MOVE

If you were to remove all outside noise, you could actually hear the sound of your own eyes moving.

In extremely quiet environments, like specialized soundproof rooms called **anechoic chambers**, people have reported hearing strange internal sounds—like blood flowing, joints creaking, and yes, even the **muscles moving their eyes**. It's a reminder that your body is never truly silent... even when the world around you is.

Mind-Blowing Human Body Fact #33

YOUR FINGERNAILS REVEAL YOUR HEALTH

Your fingernails grow about **3 millimeters per month**, but they also double as tiny health indicators.

Changes in color, texture, or shape can signal underlying issues—from vitamin deficiencies to circulation problems and even heart or liver conditions. Ever noticed white spots or vertical ridges? Those aren't just cosmetic quirks—they're clues your body leaves behind like a medical diary, written in keratin.

Mind-Blowing Human Body Fact #34

YOU REACT BEFORE YOU THINK

Your body can respond to danger **before your brain even processes it**.

Reflexes like pulling your hand away from a hot surface happen through a lightning-fast loop between your spinal cord and muscles— **bypassing your brain entirely**. That's why you can move before you even feel the pain. It's your nervous system acting like an emergency autopilot, protecting you in the split-second gap between sensation and awareness.

Mind-Blowing Human Body Fact #35

YOUR BODY HAS BUILT-IN NIGHT VISION

Your eyes adjust to darkness using a substance called **rhodopsin**, which builds up in your retina in low light.

It takes about **20 to 30 minutes** for your eyes to fully adapt to the dark, as rhodopsin regenerates and boosts your night vision. This is why stars slowly appear after you've been outside for a while at night. Fun twist? Rhodopsin is so light-sensitive that even a quick flash can wipe it out—sending your eyes back to square one.

Mind-Blowing Human Body Fact #36

YOUR BRAIN CAN REWIRE ITSELF

Your brain isn't fixed—it's constantly changing, adapting, and rewiring based on your experiences.

This incredible ability is called **neuroplasticity**, and it allows you to learn new skills, form memories, and even recover from injury. Whether you're mastering an instrument or just building a new habit, your brain is reshaping its own circuitry to support that growth. It's like having an internal upgrade system that never stops running.

Mind-Blowing Human Body Fact #37

YOU'VE GOT INVISIBLE MITES ON YOU

Right now, there are **tiny mites living on your skin**—especially around your eyelashes and nose.

Called **Demodex**, these microscopic creatures are completely harmless and actually quite common. They feed on dead skin cells and oils, and most people host them without ever knowing. They come out mostly at night, crawling slowly across your skin... but don't worry—they've been with you your whole life, and you're no worse for it.

Mind-Blowing Human Body Fact #38

YOU CAN SMELL IN STEREO

Your two nostrils don't just double your airflow—they help you **locate the direction of a smell**.

Each nostril picks up scents slightly differently and at slightly different times. Your brain compares these inputs—just like it does with your ears for sound—to pinpoint where an odor is coming from. It's called **binaral smelling,** and it gives your nose a subtle but powerful sense of spatial awareness.

Mind-Blowing Human Body Fact #39

YOU'RE MOSTLY MICROBIAL

For every human cell in your body, there are about **as many — if not more — microbial cells** living on and inside you.

These bacteria, viruses, fungi, and other microorganisms form your **microbiome**, playing essential roles in digestion, immunity, and even mental health. In fact, your gut alone contains **trillions** of microbes — more than the number of stars in our galaxy. So in a way, you're less of a single organism… and more of a bustling biological ecosystem.

Mind-Blowing Human Body Fact #40

YOU BREATHE THROUGH ONE NOSTRIL AT A TIME

At any given moment, you're mostly breathing through **just one nostril**—and it switches every few hours.

This is called the **nasal cycle**, a natural process where the tissue inside one nostril swells slightly while the other opens up. The roles slowly reverse throughout the day, helping to regulate airflow, filter particles, and keep your sense of smell sharp. It's like your nose is taking turns to stay in top form... and you probably never even noticed.

Mind-Blowing Human Body Fact #41

YOU HAVE A DOMINANT EYE

Just like being right- or left-handed, most people have a **dominant eye** that their brain favors for processing visual input.

This preferred eye does more of the heavy lifting when it comes to focusing, depth perception, and aiming—especially during activities like photography, archery, or even looking through a peephole. Interestingly, your dominant eye doesn't always match your dominant hand. It's another quirky reminder of how uniquely wired each body really is.

Mind-Blowing Human Body Fact #42

YOUR VOICE IS ONE OF A KIND

Your voice is as unique as your fingerprint—no two are exactly alike.

It's shaped by the size and shape of your **vocal cords, throat, mouth**, and **nasal passages**, creating a distinct sound signature. Even identical twins have slightly different voices. That's why voice recognition technology is becoming more common—your voice is a biometric ID you carry everywhere, whether you're speaking, singing, or shouting from the rooftops.

Mind-Blowing Human Body Fact #43

YOUR BODY GLOWS WHEN YOU EXERCISE

During intense workouts, your muscles produce **a faint blue glow**—but it's completely invisible to the naked eye.

This glow comes from **free radicals** interacting with proteins in your body, releasing tiny flashes of **biophotons**—particles of light emitted by living tissue. Scientists have captured this phenomenon using ultra-sensitive cameras. So technically, when you break a sweat, you're not just burning calories... you're literally lighting up.

Mind-Blowing Human Body Fact #44

YOUR BRAIN WORKS HARDER WHEN YOU DREAM

During REM sleep—the stage when you dream—your brain can be **even more active than when you're awake**.

It's processing emotions, sorting memories, solving problems, and sometimes sparking wild imagination. Brain scans show intense activity in areas linked to vision, emotion, and creativity. That's why dreams can feel vivid, bizarre, or deeply meaningful. So while your body rests, your brain is pulling the night shift, running a high-powered dream machine.

Mind-Blowing Human Body Fact #45

YOUR FEET HAVE A QUARTER OF YOUR BONES

Each foot contains **26 bones**, and together they account for **52 of the 206 bones** in your entire body.

That's about **25% of your total bone count**, packed into two relatively small spaces. Add in **33 joints** and over **100 muscles, tendons, and ligaments**, and your feet are engineering masterpieces—designed to support your weight, absorb shock, and keep you balanced. Every step is a biomechanical symphony playing beneath you.

Mind-Blowing Human Body Fact #46

YOU CAN SURVIVE WITH HALF A BRAIN

In rare cases, people have lived full lives after having **half of their brain surgically removed**—a procedure called **hemispherectomy**.

Usually done to treat severe epilepsy, this operation sounds impossible, but the brain's remarkable **plasticity** allows the remaining half to **take over lost functions**. Especially in children, the brain can rewire itself to compensate for what's missing. It's one of the most jaw-dropping examples of human adaptability in all of medicine.

Mind-Blowing Human Body Fact #47

YOUR EYES CAN DETECT A CANDLE MILES AWAY

Under the right conditions, the human eye can see a **single candle flame** from as far as **1.6 miles** away.

That's thanks to the incredible sensitivity of the **rods** in your retina, which can respond to just a handful of photons—the smallest units of light. While you'd need total darkness and no obstructions, this fun fact proves your eyes are far more capable than they seem. They're not just windows to the soul—they're precision optical instruments.

Mind-Blowing Human Body Fact #48

YOUR SKIN CAN SMELL

Believe it or not, your skin has **olfactory receptors**—the same kind used by your nose to detect scents.

These receptors are especially concentrated in areas like your **hair follicles**, where they can respond to chemical signals. Research has shown that skin cells can "smell" certain compounds and even respond by **healing faster** or growing more efficiently. So in a strange but literal sense, your skin can sniff out what's happening around it.

Mind-Blowing Human Body Fact #49

YOUR BODY CAN MAKE ITS OWN WATER

When you metabolize food, your body actually **creates water** as a byproduct.

This process, called **metabolic water production**, happens when your cells break down fats, carbs, and proteins for energy. Astronauts and desert animals like camels rely heavily on this internal water supply to survive in extreme conditions. So even if you're not sipping from a bottle, your body is quietly hydrating itself at the cellular level.

Mind-Blowing Human Body Fact #50

YOU'RE TALLER IN SPACE

Without gravity compressing your spine, astronauts grow up to **2 inches taller** while in space.

On Earth, gravity constantly squishes the soft discs between your vertebrae. But in the microgravity of space, those discs expand, and the spine stretches out—making astronauts slightly taller during their missions. The effect reverses once they return to Earth, but for a while, they get to enjoy a temporary height boost... and maybe a better reach on the top shelf.

Mind-Blowing Human Body Fact #51

YOUR KIDNEYS FILTER 50 GALLONS A DAY

Your kidneys filter your entire blood supply **about 50 times per day**, processing around **50 gallons of fluid** in just 24 hours.

These two fist-sized organs work quietly behind the scenes, removing waste, balancing fluids, regulating blood pressure, and even helping make red blood cells. Despite their size, they're absolute workhorses — tiny purification plants keeping your internal environment in perfect balance.

Mind-Blowing Human Body Fact #52

YOUR BODY HAS BACKUP ORGANS

Incredibly, the human body includes **dupli-cate organs** that allow you to live a normal life even if one stops working.

You only need **one kidney**, **one lung**, and even part of a **liver** to survive. In some cases, people have donated half their liver or a kidney and continued living completely healthy lives. It's nature's built-in redundancy system—like having a spare tire, but for your most vital functions.

Mind-Blowing Human Body Fact #53

YOU HAVE A NATURAL DETOX SYSTEM

Forget juice cleanses—your body detoxes itself **every single day** through your **liver, kidneys, lungs, skin, and intestines**.

These organs work together to break down toxins, filter waste, and eliminate harmful substances through urine, sweat, breath, and bowel movements. Your liver, especially, is a detox superhero—processing everything from medications to alcohol. So while trendy detox fads come and go, your body's internal cleaning crew is always on the job.

Mind-Blowing Human Body Fact #54

YOUR BRAIN HAS NO PAIN RECEPTORS

The brain itself can't feel pain—even though it processes all pain signals in the body.

This is why surgeons can perform brain surgery on patients who are **wide awake**, with only local anesthesia for the scalp. The surrounding tissues can hurt, but the brain? Totally numb. It's a strange twist: the organ responsible for *feeling* pain is completely immune to it.

Mind-Blowing Human Body Fact #55

YOUR BLOOD TYPE CAN AFFECT YOUR HEALTH

Your blood type does more than determine who you can donate to—it may also influence your **risk for certain diseases**.

Studies have linked different blood types to variations in **heart disease, clotting disorders, and even stomach bugs**. For example, people with type O blood may have a lower risk of heart attack, while those with type A might be more prone to certain infections. It's a subtle but fascinating layer of how your biology shapes your health.

Mind-Blowing Human Body Fact #56

YOUR TEETH ARE AS HARD AS STEEL

Tooth enamel—the outer layer of your teeth—is the **hardest substance in the human body**.

It's even harder than bone and ranks just below steel on the hardness scale. This tough shell protects your teeth from daily wear, chewing, and even acid attacks. But once it's damaged, it **can't regenerate**, which is why dentists are always stressing good oral hygiene. You've got steel-strength gear in your mouth—treat it like treasure.

Mind-Blowing Human Body Fact #57

YOU CAN SURVIVE WITHOUT A PULSE

Thanks to modern medical devices, it's possible to live **without a detectable heartbeat**.

Some advanced heart pumps, like **LVADs (Left Ventricular Assist Devices)**, create continuous blood flow instead of the typical pulsing beat. Patients with these implants have no pulse at all—yet remain fully conscious and functional. It's a medical marvel that challenges our very definition of what it means to be alive.

Mind-Blowing Human Body Fact #58

YOU CAN HEAR BETTER WHEN YOU YAWN

Yawning doesn't just help you wake up—it also **opens your Eustachian tubes**, equalizing pressure in your ears.

This small adjustment can temporarily **improve your hearing,** especially in situations like takeoff or landing in an airplane. It's your body's natural way of tuning its internal sound system, clearing out that muffled feeling and letting the vibrations flow freely. So next time you yawn, you're not just stretching—you're subtly sharpening your senses.

Mind-Blowing Human Body Fact #59

YOUR BODY EMITS A MAGNETIC FIELD

Your heart generates an electrical current with every beat—and that current produces a **measurable magnetic field**.

This field can be detected several feet away using sensitive instruments like **magnetocardiograms**. It's one of the most powerful electromagnetic fields produced by the body, and researchers are studying how it may play a role in everything from emotional communication to heart health. You're not just alive—you're subtly magnetic.

Mind-Blowing Human Body Fact #60

YOUR BRAIN RUNS ON 20 WATTS

Your brain uses about **20 watts of power**— roughly the same as a light bulb.

That's all it takes to manage your thoughts, memories, emotions, movements, and more. Despite making up only about **2% of your body weight**, the brain gobbles up **20% of your energy**. It's an ultra-efficient supercomputer running constantly on just a trickle of electricity. Bright ideas, indeed.

Mind-Blowing Human Body Fact #61

YOUR BODY CLOCK ISN'T 24 HOURS

Your internal body clock—called the **circadian rhythm**—actually runs on a cycle of about **24.2 hours**, not exactly 24.

Left without exposure to natural light, like in cave experiments, people tend to drift slightly later each day. That's why sunrise and sunset are so important—they help reset and sync your internal clock with the Earth's rotation. You're basically running your own little time zone... and nature keeps you on schedule.

Mind-Blowing Human Body Fact #62

YOUR REFLEXES ARE FASTER THAN YOU THINK

Some of your reflexes are so fast, they bypass conscious thought entirely—clocking in at just **0.1 seconds**.

That's faster than a blink and way quicker than most people can react on purpose. These lightning-fast responses are managed by your **spinal cord**, which can trigger actions like jerking your hand from a hot stove before your brain even registers the pain. It's your body's built-in emergency response system—faster than fear, quicker than thought.

Mind-Blowing Human Body Fact #63

YOUR BELLY BUTTON HOLDS A MICROBIOME

That little dimple in your stomach is home to a surprisingly complex **ecosystem of bacteria**—some species found **nowhere else** on your body.

In one study, scientists discovered over **2,300 types** of microbes living in belly buttons, including rare organisms previously seen only in polar ice caps and deep-sea vents. Each person's navel biome is unique, like a microbial fingerprint tucked away in plain sight.

Mind-Blowing Human Body Fact #64

YOU GROW MORE IN THE SUMMER

Studies have shown that people tend to **grow slightly faster during the summer months**.

This is likely due to a mix of **increased sunlight**, which boosts **vitamin D** production, and more physical activity, which stimulates bone growth—especially in kids and teens. Your body responds to the changing seasons in subtle ways, and summer just might be your biological growth spurt season.

Mind-Blowing Human Body Fact #65

YOUR SWEAT HAS NO SMELL

Sweat itself is **odorless**—it only starts to smell when it mixes with bacteria on your skin.

Your body releases two types of sweat: one for cooling down (mostly water and salt), and another from **apocrine glands**, which contains proteins and fats. It's this second type that bacteria love to feast on—creating body odor as a byproduct. So the smell? It's not really sweat... it's the party bacteria throw afterward.

Mind-Blowing Human Body Fact #66

YOUR HANDS HAVE A BRAIN BIAS

Each hand is controlled by the **opposite hemisphere** of your brain—your right hand by the left brain, and your left hand by the right.

This "crossed wiring" is why right-handed people often excel at logic and language (left-brain functions), while lefties sometimes show strengths in creativity and spatial reasoning (right-brain functions). It's also why brain injuries on one side can affect movement on the other. Your brain is running a constant game of mirrored control—like a puppeteer working from across the stage.

Mind-Blowing Human Body Fact #67

YOUR BODY HAS HIDDEN SYMMETRY

While your organs may look uneven on the outside, your body is built with a surprising amount of **internal symmetry**.

From the left and right hemispheres of your brain to the paired structure of kidneys, lungs, and limbs, your body is designed with **bilateral balance** in mind. Even your face—though not perfectly symmetrical—follows this mirrored layout. It's a subtle architectural principle that helps with movement, coordination, and even beauty perception.

Mind-Blowing Human Body Fact #68

YOU CAN CRY WITHOUT TEARS

Not all crying involves tears—your body can produce **emotional crying** without any liquid at all.

Tearless crying often happens under intense emotional stress or in people with certain medical conditions. And interestingly, **not all animals** cry emotionally the way humans do. While other species produce tears to protect their eyes, humans are uniquely wired to shed them from sadness, joy, or even laughter. So when your eyes stay dry, your emotions may still be in full flood.

Mind-Blowing Human Body Fact #69

YOU HAD GILLS IN THE WOMB

During early development, human embryos form **gill-like structures**—a throwback to our ancient aquatic ancestry.

These structures, called **pharyngeal arches**, later evolve into parts of the jaw, ears, and throat. But for a brief time, you looked a bit like a tiny fish, echoing the shared blueprint of all vertebrates. It's a surreal reminder that your body carries the evolutionary story of life itself—right from the womb.

Mind-Blowing Human Body Fact #70

YOUR HAIR REMEMBERS YOUR HISTORY

Each strand of hair carries a **chemical record of your past**, including diet, stress levels, and even drug use.

Hair grows about **half an inch per month**, and as it does, it locks in molecular traces of what's circulating in your bloodstream. Forensic scientists and doctors can analyze these strands like biological time capsules—revealing what was happening in your body **weeks or even months ago**. Your hair isn't just style—it's science in filament form.

Mind-Blowing Human Body Fact #71

YOU LAUGH BEFORE YOU UNDERSTAND

Laughter often kicks in **before your brain fully processes a joke**.

That's because laughter is a **primal social reflex**, deeply rooted in the brain's older structures—like the amygdala and brainstem. It evolved as a bonding tool, not just a response to humor. In fact, studies show we're more likely to laugh in groups than when alone, even if something's equally funny. So sometimes, your body decides to laugh... before your mind gets the memo.

Mind-Blowing Human Body Fact #72

YOU DREAM EVERY NIGHT—EVEN IF YOU FORGET

Whether you remember them or not, you experience **multiple dreams every night**—usually during **REM sleep**.

Most dreams last between **5 to 20 minutes**, and the average person has around **4 to 6** per night. You just tend to forget them **within minutes** of waking up unless something emotional or unusual sticks. So even if you swear you "never dream," your brain's been quietly making movies while you sleep.

Mind-Blowing Human Body Fact #73

YOUR SPINE IS A SHOCK ABSORBER

Your spine isn't just a stack of bones—it's a **flexible suspension system** designed to absorb impact.

Between each vertebra are **gel-filled discs** that cushion your movements, protect your nerves, and let you bend, twist, and stretch. These discs even **compress slightly** throughout the day, which is one reason you're shorter at night. It's a marvel of biological engineering— strong enough to support you, soft enough to protect you.

Mind-Blowing Human Body Fact #74

YOU'RE BIOLUMINESCENT AT BIRTH

Newborn babies emit a faint **ultraviolet glow**—a natural phenomenon caused by **bilirubin**, a pigment produced as their bodies break down old red blood cells.

While invisible to the naked eye, this glow can be detected with special equipment and is often used in hospitals to monitor newborns for **jaundice**. It's one of the rare times in life when your body literally lights up as part of the transition into the world.

Mind-Blowing Human Body Fact #75

YOUR BONES ARE ALWAYS REMODELING

Even when you're fully grown, your bones are **constantly breaking down and re-building themselves**.

This process, called **bone remodeling**, re-places old or damaged bone tissue with fresh material at a rate of about **10% per year**. Spe-cialized cells called **osteoclasts** and **osteoblasts** handle the demolition and construction, keep-ing your skeleton strong and responsive to stress. Your bones aren't static—they're a liv-ing, dynamic scaffold in constant flux.

Mind-Blowing Human Body Fact #76

YOU HAVE TASTE BUDS IN UNEXPECTED PLACES

While most of your taste buds are on your tongue, some are also located on the **roof of your mouth**, **throat**, and even your **esophagus**.

These scattered receptors help round out your sense of taste and are part of why food feels so flavorful beyond just your tongue. It's a full-mouth experience—literally! Your body built a backup flavor system, just in case your tongue ever gets lazy.

Mind-Blowing Human Body Fact #77

YOU CAN SMELL FEELINGS

Humans can actually **detect emotional states** like fear or happiness through scent—without even realizing it.

When you're scared or anxious, your body releases subtle chemical signals in sweat called **chemosignals**. Studies show that people exposed to "fear sweat" unconsciously display more alert or cautious behavior. It's a quiet, primal form of communication—your body broadcasting your mood without a single word.

Mind-Blowing Human Body Fact #78

YOU BLINK IN PERFECT SYNC

Your eyelids **blink simultaneously with milliseconds precision**, thanks to a finely tuned neural network coordinating both eyes.

Even though they're controlled by separate muscles, your brain ensures your eyelids close and open at the same exact moment—every time. It's one of those tiny, seamless feats your body pulls off constantly without you ever noticing. Perfect symmetry, hundreds of times a day.

Mind-Blowing Human Body Fact #79

YOUR FINGERS DON'T HAVE MUSCLES

Despite their strength and dexterity, your fingers contain **no muscles at all**.

All the motion comes from **tendons** pulled by muscles located in your **forearms and palms**. It's like a system of biological puppet strings, with every tap, grip, and snap controlled remotely. So next time you wiggle your fingers, thank your forearms—they're doing the heavy lifting behind the curtain.

Mind-Blowing Human Body Fact #80

YOU HAD MORE NEURONS AS A BABY

At birth, your brain had **more neurons than it does now**—around **100 billion**, the highest it will ever be.

As you grow, your brain undergoes a process called **synaptic pruning**, where unused connections are trimmed to make the system more efficient. It's kind of like decluttering a messy closet—fewer items, but better organized. So while you technically lose brain cells, you gain processing power.

Mind-Blowing Human Body Fact #81

YOUR BODY CAN PREDICT THE WEATHER

Some people really **can feel a storm coming**—thanks to how the body responds to changes in **barometric pressure**.

As atmospheric pressure drops before a storm, tissues in your body can expand slightly, triggering **joint pain**, **headaches**, or sinus pressure. It's especially noticeable in people with arthritis. So when grandma says she feels rain in her knees… she's not just being poetic—she's being physiological.

Mind-Blowing Human Body Fact #82

YOU MAKE NOISE WHEN YOU DIGEST

That rumbling in your stomach? It has a name: **borborygmus**—and it's totally normal.

It happens when your intestines move gas and fluids during digestion, even when you're not hungry. In fact, your gut makes these sounds **all day long**, but they're louder when your stomach is empty because there's less material to muffle the noise. So the next time your belly "talks," just know—it's busy at work.

Mind-Blowing Human Body Fact #83

YOU CAN HEAR THROUGH YOUR BONES

Sound doesn't just travel through your ears—it can also reach your inner ear **through the bones in your skull.**

This is called **bone conduction**, and it's how you can hear your own voice differently than others do. It's also the principle behind certain types of hearing aids and headphones, which bypass the eardrum entirely. Your skeleton isn't just structural—it's part of your audio system too.

Mind-Blowing Human Body Fact #84

YOUR PUPILS REACT TO EMOTION

Your pupils don't just respond to light—they also **change size based on how you feel**.

When you're excited, attracted to someone, or deep in concentration, your pupils can **dilate**, letting in more visual information. It's an ancient survival mechanism tied to heightened alertness—and it happens so subtly, most people never notice. Your eyes really are windows to your emotional state.

Mind-Blowing Human Body Fact #85

YOU HEAL FASTER WHEN YOU'RE HAPPY

Positive emotions can actually **speed up the healing process**.

Studies have shown that people who are optimistic or in a good mood **recover faster from injuries and surgeries**. That's because happiness can lower stress hormones and boost immune function—giving your body a better environment to repair itself. So yes, laughter and joy truly are good medicine.

Mind-Blowing Human Body Fact #86

YOU CAN TASTE WORDS AND SOUNDS

Some people experience **synesthesia**, a rare condition where senses blend—like **tasting shapes** or **hearing colors**.

One of the most fascinating forms is **lexical-gustatory synesthesia**, where certain words or sounds trigger specific taste sensations. For example, the word "Monday" might taste like mint, or a violin note could taste like chocolate. It's not imagination—it's the brain's sensory wires crossing in extraordinary ways.

Mind-Blowing Human Body Fact #87

YOUR EYELASHES HAVE A LIFESPAN

Each of your eyelashes lives for about **3 to 5 months** before falling out and being replaced.

They grow in cycles, just like the hair on your head, and protect your eyes from dust, debris, and light. On average, you have **75 to 150 lashes** on your upper eyelid alone—and even though they seem delicate, they're remarkably resilient. Your lashes are tiny bodyguards with a built-in refresh schedule.

Mind-Blowing Human Body Fact #88

YOUR TONGUE NEVER RESTS

Even when you're not eating or talking, your tongue is constantly **shifting, adjusting, and helping with swallowing**.

It moves to keep saliva flowing, clears food particles, and maintains your airway while you sleep. Over the course of a day, your tongue makes **thousands of tiny, unconscious movements**—it's one of the most active muscles in your body, and it never truly clocks out.

Mind-Blowing Human Body Fact #89

YOU SHED DNA EVERYWHERE YOU GO

Every time you touch something, you leave behind tiny traces of your **genetic fingerprint**—in the form of skin cells, sweat, and hair.

This constant shedding means your DNA can be found on doorknobs, keyboards, clothing, and even in the air around you. Forensic scientists call it **touch DNA**, and it's powerful enough to identify someone from just a few skin cells. You're unknowingly leaving a genetic trail wherever you go.

Mind-Blowing Human Body Fact #90

YOUR VOICE CHANGES AS YOU AGE

As you get older, your voice naturally **deepens, softens, or weakens** due to changes in your vocal cords and respiratory system.

In children, voices are higher-pitched because their vocal cords are shorter and thinner. During puberty, hormonal changes cause the cords to lengthen—especially in boys—leading to that infamous "voice crack." Over time, aging can dry out or stiffen the cords, gradually altering tone and volume. Your voice, like your face, tells the story of time.

Mind-Blowing Human Body Fact #91

YOU HAVE A DOMINANT NOSTRIL

Just like being right- or left-handed, most people have a **dominant nostril** they rely on more for breathing.

Over time, your body subtly favors one nostril due to the **nasal cycle**—a natural rhythm where airflow alternates between sides every few hours. But even outside that cycle, one nostril usually handles slightly more airflow than the other. So while you breathe through both, you might unknowingly be a "right-nose" or "left-nose" breather.

Mind-Blowing Human Body Fact #92

YOUR BRAIN STARTS SLOWING AT 24

Research shows that certain **cognitive functions**, like reaction time and processing speed, begin a **gradual decline around age 24**.

But don't panic—other brain skills, like **pattern recognition, vocabulary, and emotional intelligence**, often improve well into middle age and beyond. It's less about decline and more about shifting strengths. Your brain's always adapting, just in different ways at different stages.

Mind-Blowing Human Body Fact #93

YOUR FINGERS PRUNE FOR GRIP

Those wrinkly fingertips in the bath? They're not just soggy skin—they're part of a built-in **grip-enhancement system**.

When submerged, your nervous system triggers blood vessels in your fingers to constrict, creating wrinkles. Researchers believe this improves your ability to **grip wet or slippery objects**, kind of like tire treads on a rainy road. It's your body adapting in real-time for better performance.

Mind-Blowing Human Body Fact #94

YOUR EYES CAN SPOT A SINGLE PHOTON

In pitch-black conditions, the human eye is sensitive enough to detect **just one photon**— the smallest unit of light.

It doesn't happen often, but under perfect circumstances, a single photon hitting your retina can trigger a visual response. That's an **astonishing level of sensitivity**, showcasing just how finely tuned your vision truly is. Even the tiniest flicker in the darkness doesn't go unnoticed.

Mind-Blowing Human Body Fact #95

YOU CAN SURVIVE WITHOUT 90% OF YOUR LIVER

The liver is the only organ that can **regenerate from just 10% of its original mass**.

While we've touched on its regrowth ability before, here's the jaw-dropper: **you can survive and rebuild full liver function with as little as one-tenth of the organ remaining**. That's why partial liver transplants are possible—and successful. It's the closest thing your body has to a biological superpower.

Mind-Blowing Human Body Fact #96

YOUR BONES REACT TO GRAVITY

Your skeleton isn't just structural—it's **constantly responding to gravity and pressure** to stay strong.

Astronauts in space **lose bone density** due to the lack of gravitational stress, sometimes up to **1% per month**. On Earth, weight-bearing activities like walking or lifting keep your bones dense and healthy. In short, your bones literally need gravity to thrive—it's one of the quiet forces shaping your frame every day.

Mind-Blowing Human Body Fact #97

YOU HAVE A BUILT-IN COOLING SYSTEM

Your body can produce up to **1.5 liters of sweat per hour** during intense activity—part of an advanced temperature regulation system.

Sweating isn't just about moisture—it cools your skin as the sweat evaporates, preventing overheating. What's wild is how efficient this system is: humans are among the **best sweaters in the animal kingdom**, which helped our ancestors hunt by outlasting overheated prey. You're basically a self-cooling endurance machine.

Mind-Blowing Human Body Fact #98

YOU CARRY ELECTRICITY IN YOUR MUSCLES

Your muscles move because of tiny **electrical signals** generated by nerve impulses—and these signals can be measured.

Devices like **EMGs (electromyograms)** detect this electrical activity, which is what powers everything from your heartbeat to a simple smile. In fact, these signals are strong enough to control **robotic limbs** in prosthetic technology. So every flex, twitch, and blink is powered by your body's own electrical grid.

Mind-Blowing Human Body Fact #99

YOU SHED YOUR ENTIRE SKIN EACH MONTH

Your skin completely renews itself rough-ly **every 27 days**, replacing old cells with fresh ones.

That means in a single year, you go through about **13 full cycles of skin regeneration**. Most of the dead skin cells you shed—**millions per day**—end up as household dust. It's like your body is constantly giving itself a microscopic makeover... without you ever noticing.

Mind-Blowing Human Body Fact #100

YOU'RE MADE OF STARDUST

Every atom in your body — carbon, calcium, iron, and more — was **forged in the core of ancient stars** that exploded in supernovas billions of years ago.

Those scattered elements eventually formed planets, plants, and people. So when scientists say "you're made of stardust," they mean it literally. The universe didn't just create you... **you are the universe, thinking about itself**.

CONCLUSION

Congratulations! You've just explored *100 Mind-Blowing Human Body Facts* and uncovered some of the most astonishing secrets hiding beneath your skin. From invisible glows to backup organs, from reflexes faster than thought to bones that rebuild themselves—you've seen just how strange, surprising, and spectacular your body truly is.

But here's the thing about the human body—it's still full of mysteries. For every fact you've read, there are countless more being discovered every day. Scientists are constantly unlocking new layers of how we work, evolve, and adapt. Maybe this book has sparked your curiosity, or maybe it's simply reminded you how wild it is to be human.

The truth is, you don't need a microscope or a lab coat to be amazed by biology. All it takes is a sense of wonder and a willingness to ask, "How does that even work?" Because the body isn't just a machine—it's a living story that never stops unfolding.

So as you close this book, don't think of it as the end. Think of it as your first deep breath into a world that's more incredible than science fiction—and it's happening inside you, right now.

Until next time, stay curious, stay amazed, and remember: the most mind-blowing story you'll ever read… is you.

ACKNOWLEDGEMENTS

Creating *100 Mind-Blowing Human Body Facts* has been a journey filled with curiosity, caffeine, and more than a few late-night "wait… is that true?!" moments. While my name may be on the cover, this book wouldn't exist without the inspiration, encouragement, and enthusiasm of some incredible people.

First, a huge thank you to the scientists, researchers, educators, and medical professionals whose work continues to uncover the wild and wonderful mysteries of the human body. Your discoveries and dedication are the backbone of this book—and the reason we all get to say, "I had no idea my body could do that."

To my family and friends, thank you for letting me geek out over things like skin cells, neurons, and sneaky biological superpowers. Your support (and your ability to smile through endless fun facts about bones and blood vessels) made this process a whole lot more enjoyable.

To my readers—thank *you*. Whether you're here for the trivia, the weird science, or just a fun reminder of how miraculous your body really is, I'm so glad you picked up this book. Your curiosity is what keeps the world of science exciting and alive.

And finally, to the human body itself—thank you for being the most bizarre, brilliant, and endlessly fascinating subject imaginable. You've given us 100 mind-blowing facts... and countless more waiting to be discovered.

Here's to wonder, to science, and to this incredible machine we all call home.

ABOUT THE AUTHOR

Felix Grayson is a storyteller at heart, driven by an insatiable curiosity for the strange, surprising, and downright unbelievable things that happen inside the human body. With a passion for uncovering the world's weirdest facts and making science fun for everyone, Felix created *100 Mind-Blowing Human Body Facts* to entertain, amaze, and spark a deeper appreciation for the incredible machine we all live in.

When he's not buried in biology books or chasing down mind-bending medical trivia, Felix enjoys exploring science museums, getting lost in weird health documentaries, and pondering life's most fascinating questions over a strong cup of coffee. A firm believer that learning should be fun (and a little weird), Felix invites you to marvel at the science of *you*—proving that the human body might just be the most mind-blowing subject of all.

www.ingramcontent.com/pod-product-compliance
Lightning Source LLC
Chambersburg PA
CBHW031850200326
41597CB00012B/347